VIET NAM'S ECOSYSTEM FOR TECHNOLOGY STARTUPS

Truong Thinh Pham and Aimee Hampel-Milagrosa

JULY 2022

Country Report No. 4
Ecosystems for Technology Startups in Asia and the Pacific

AN DEVELOPMENT BANK

ADB

© 2022 Asian Development Bank
6 ADB Avenue, Mandaluyong City, 1550 Metro Manila, Philippines
Tel +63 2 8632 4444; Fax +63 2 8636 2444
www.adb.org

Some rights reserved. Published in 2022.

ISBN 978-92-9269-630-6 (print); 978-92-9269-631-3 (electronic); 978-92-9269-632-0 (ebook)
Publication Stock No. TCS220294-2
DOI: http://dx.doi.org/10.22617/TCS220294-2

The views expressed in this publication are those of the authors and do not necessarily reflect the views and policies of the Asian Development Bank (ADB) or its Board of Governors or the governments they represent.

ADB does not guarantee the accuracy of the data included in this publication and accepts no responsibility for any consequence of their use. The mention of specific companies or products of manufacturers does not imply that they are endorsed or recommended by ADB in preference to others of a similar nature that are not mentioned.

By making any designation of or reference to a particular territory or geographic area, or by using the term "country" in this document, ADB does not intend to make any judgments as to the legal or other status of any territory or area.

Corrigenda to ADB publications may be found at http://www.adb.org/publications/corrigenda.

Notes:
In this publication, "$" refers to United States dollars and S$ refers to Singapore dollars.
ADB recognizes "China" as the People's Republic of China and "Vietnam" as Viet Nam, "Danang" as Da Nang, and "Hanoi" as Ha Noi.

All photos are owned by ADB unless otherwise stated.

Cover design by Joe Mark Ganaban.

Contents

Tables and Figures

Foreword

Technology-based startup enterprises—or tech startups—are an increasingly important part of the business landscape in Asia and the Pacific. These enterprises use new technologies to create new products or services, or to provide services in a new way. Most startups will not survive, but some will succeed and make an important contribution to economic development. Tech companies like Facebook, Google, and Amazon are among the largest companies in the world today, and tech companies like Tencent, Gojek, Grab, VNG, VnPay, and MoMo are also among the leading emerging companies in Asia. The technology and dynamism they provide are important for economic growth.

Tech startups do not grow in a vacuum. They need access to funding, often from venture capitalists; skilled personnel, including experts in technology and business; good digital infrastructure; and supportive government policies. A strong ecosystem is critical for turning new ideas into commercially viable businesses. Given their growing importance, it is important to better understand the ecosystem in which tech startups develop.

This report assesses the state of tech startups in Viet Nam with a focus on the ecosystem. It examines the extent to which the system supports the growing number of startups in the country. The report focuses on two sectors: agritech and healthtech. While fintech and e-commerce startups are the most prevalent in Viet Nam and other countries, these two sectors were chosen because startups in these sectors not only become successful businesses, but also have a strong impact on development. They support human capital formation and impact the largely poor rural sector.

The report provides recommendations on how the government and other stakeholders can strengthen the ecosystem to enable tech startups to flourish in Viet Nam. It is one of a series of ADB reports that studies the ecosystem of tech startups in Asia and the Pacific.

Albert Park
Chief Economist
Asian Development Bank

Acknowledgments

This report was prepared by Truong Thinh Pham and Aimee Hampel-Milagrosa. The latter, along with Paul Vandenberg and Matthias Helble, guided the research project. Rana Hasan and Lei Lei Song provided overall management support. The Asian Development Bank's Viet Nam Resident Mission reviewed the report and solicited comments from the Government of Viet Nam.

The authors would like to thank key informants from ministries, incubators, accelerators, development partners, investors, academic institutions, and startups who provided invaluable insights that were indispensable for the preparation of the study. The draft report was reviewed by the Department of Market Development and Science and Technology Enterprises, Department of Technology Application and Development, and Supporting the National Innovative Startup Ecosystem to 2025 project, all from the Ministry of Science and Technology (MOST), which provided extensive comments and suggestions. Their involvement is highly appreciated and has helped provide a more detailed overview of the government's tech startup activities in Viet Nam, especially since the outbreak of the coronavirus disease (COVID-19) pandemic. Tuesday Soriano copyedited the report, and Amanda Isabel Mamon provided administrative support, contracting, and manuscript management.

Abbreviations

HCMC	Ho Chi Minh City
IPR	intellectual property rights
MOST	Ministry of Science and Technology
NATEC	National Agency for Technology Entrepreneurship and Commercialization Development
NIC	National Innovation Center
SATI	State Agency for Technology Innovation
SMEs	small and medium-sized enterprises
SME Law	Law on Small and Medium Enterprise Support
STEM	Science Technology Engineering and Mathematics
US	United States
VC	venture capital
VCCI	Vietnam Chamber of Commerce and Industry
VCIC	Vietnam Climate Innovation Center
VIISA	Vietnam Startup Acceleration Fund
VMI	Vietnam Mentors Initiative

Executive Summary

In Viet Nam, the concept of technology-based startups is relatively new as it is only in 2016 when all the components of a startup ecosystem became available. There are no official figures on the number of tech startups, but VN Express, a leading Vietnamese international online newspaper, estimates the total number in 2021 to be around 3,800, with four homegrown unicorns (companies worth over $1 billion): VNG, VnPay, SkyMavis, and MoMo. Viet Nam attracted $448 million in capital investment in startups in 2018, doubling to $861 million in 2019. After the pandemic, investments in Vietnamese tech startups reached a record high of $1.4 billion (NIC 2021).

The national startup ecosystem is focused on developing vibrant, fast-growing businesses based on the use of intellectual property, technology, and innovative business models. Policies to encourage investment in startups in Viet Nam are gradually being completed to demonstrate the government's commitment to improving the business environment for tech startups, such as the numerous government-sponsored pitching competitions and the visit-and-learn experiences to other countries with developed startup ecosystems. The Law on Small and Medium Enterprise Support (SME Law) officially established the legal status of startups. The most important legal document to date, Decree No. 38/2018/ND-CP dated 11 March 2018, details the formation and operation of innovative startup funds. This created an important legal basis for the establishment of organizations supporting Vietnamese startups, and also laid the groundwork for the development of a capital market for innovative startups. As a result, several local investment funds have been established to provide financial support to startups. Branches of international venture capital (VC) offices, accelerators, and coworking spaces have been opened in Ha Noi and Ho Chi Minh City (HCMC).

Components of Viet Nam's Startup Ecosystem

Viet Nam's startup ecosystem, while still evolving, has the basic ingredients for tech startup growth, with the following notable characteristics:

State and government. In 2016, the Prime Minister approved Project 844, the most direct government intervention to create a favorable environment for Vietnamese startups. Overseen by the Ministry of Science and Technology (MOST), Project 844 is tied to specific targets for the development of tech startups. MOST established the National Technology Innovation Fund in 2015 to attract knowledge, organizations, entrepreneurs, and scientists to contribute to economic development and participate in the creation of startups.

Angel investors. The number of angel investors in Viet Nam is still small, but it is increasing. These are local and foreign angel investors—including relatives and friends—with small investments for pre-seed funding. Some of the active angel investors in the country are successful first-generation startup founders.

Venture capital. The number of VC funds operating in Viet Nam is quite small, and most are funded by the government. In 2019, about 40 VC funds were active. Also in 2019, Viet Nam attracted about 3% of total venture capital in Southeast Asia; this share has since increased to 19% in 2022. Corporate venture capital is also gaining traction in the country. Many established large corporates in traditional industries need to adapt to technology innovation to remain competitive and therefore are potential customers, partners, and even VC investors for tech startups. The number and value of VC transactions is slowly increasing.

Incubators and accelerators. In 2018, there were about 50 organizations supporting startups in Viet Nam, almost 50% more than in 2017. Some accelerators not only provide coworking space, but also act as venture capitalists, funding early-stage startups. Incubators and accelerators are a relatively new concept in Viet Nam, and the strong development of the tech startup community has highlighted their limited capacity.

Universities and research institutes. In Viet Nam, universities and research institutes, in cooperation with provincial authorities and central ministries, have created startup support units within universities. Universities organize startup events for students to promote the culture of entrepreneurship among young people. In addition, university-based coaches, startup advisors, and mentors provide guidance and advice to young startup entrepreneurs. There are also several inter-university advisory networks. Generally, students from their second year of study are encouraged to start their own businesses, preferably startups.

Survey of Agritech and Healthtech Startups

The study team conducted a survey of 11 Vietnamese startups, including 5 respondents from agritech and 6 respondents from healthtech. All of the startups surveyed are small. Agritech startups tend to be smaller and have an average of 20 employees, while healthtech startups have an average of 23 employees, including permanent and temporary workers. Most founders in both sectors have a bachelor's degree in science or information technology (IT) and are under 30 years old. Many of them had their technical ideas while studying at universities or technical schools, while the marketing ideas came much later. All startups typically have difficulty recruiting and maintaining human resources, although the problem is more serious in the healthtech sector. Market penetration and financial sustainability are problems for all startups.

In addition to their small size, all tech startups surveyed have limited equity. To raise initial capital, the startups use their own funds or those of family and friends or special funds offered by government agencies as the standard source of funding. All three types of funding are insufficient for expansion.

Some banks offer concessional rates to innovative SMEs, including tech startups. No distinction is made between SMEs and tech startups, as the loan terms are the same for both.

The common attitude of venture capital funds toward startups—and vice versa—is one of skepticism. According to founders that were interviewed,[a] VC firms are willing to invest, but only on the condition that the startup's profits are already large. For early-stage startups, this is a difficult condition to meet. Founders also said that they would usually have to pledge at least half of the equity to the VC firm, which reduces the founders' control over the startup.[b] As a result, many startups are reluctant to obtain venture capital. However, according to state agencies, this perception is unfounded. Tech startups should be more concerned with developing a proof of concept and a minimum viable product, which are much more important.

Growth Constraints

Agritech startups have identified the following barriers to growth: (i) weak physical (connectivity) and legal infrastructure for tech startups; (ii) lack of research in agriculture and technology; and (iii) weak foreign investor confidence. Many foreign VC firms are more interested in investing in startups in more lucrative sectors such as e-commerce, fintech, gaming, travel, or enterprise solutions. Due to the cyclical and seasonal nature of agriculture, agricultural startups often require long-term investments, especially when testing technology and prototypes, making it challenging to generate immediate returns like a traditional business. In relation to the connectivity barrier, the agritech founders' connection to information on, and about, the startup ecosystem players is weak. For agritech startups, they need support to expand their market, information to strategize how to meet market conditions and qualifications, and help to find suitable partners for developing businesses. Those are not readily available because they can only be found through intermediaries and support organizations.

Healthtech startups also cited the following constraints to growth: (i) widespread use of traditional medical services, (ii) weak physical and legal infrastructure, (iii) founders' lack of medical expertise and their dependence on medical providers for quality, and (iv) lack of financial skills. In Viet Nam, the use of traditional medical services is still popular, especially in the rural areas. In terms of infrastructure and legal barriers, the digitization of medical data in Viet Nam is slow, and the protection of medical information is strict. Hospitals still use many written records. Most of the founders are from the information and communication technology (ICT) sector, and their knowledge of health is low.

In the interviews, tech startups raised the following policy constraints: (i) the 49% foreign ownership limit; (ii) tax incentives granted based on location, industry, and production size; (iii) the differential treatment of offshore and domestic VC funds, with offshore funds restricted; (iv) the overlap of SME policies with startup policies; and (v) cumbersome intellectual property rights (IPR) procedures that do not meet the specific needs of startups.

---

[a] Some VC firms invest in companies that are making losses as long as revenues are growing fast and foreseeable breakeven or profit can be reached in future.
[b] It is noted that perhaps this experience is unique. VCs typically ask for minority stake.

Analysis and Recommendations

One of the more difficult tasks in creating a favorable ecosystem for tech startups is adapting the regulatory framework for innovative companies. The main legal provisions target SMEs, but only a few provisions (e.g., Project 844) target tech startups. Although SMEs and startups may seem similar in the initial stages, there is a clear difference between the two, which should be filtered into distinct policies. The general legal framework supporting startups needs to be revised from time to time.

VC financing remains elusive because of demand and supply constraints. This study has shown that there are no regulations for the formation and development of VC funds, which has created additional problems. The analysis shows that some policy reforms are needed to create an enabling environment for the creation of innovative enterprises, promote the growth of tech startups, and improve the availability of financing. To create an overall favorable ecosystem for startups, this study proposes the following policy recommendations:

- Remove barriers to business and cumbersome administrative procedures.

- Focus on policies and programs that encourage innovation.

- Strengthen regulations to protect intellectual property rights.

- Develop the legal framework for the establishment and operation of incubators and accelerators.

- Encourage joint activities between universities, research institutes, and the private sector.

- Evaluate the activities of incubators, accelerators, and support centers.

To create a favorable startup ecosystem for agritech and healthtech startups, the following policy recommendations are proposed:

- Develop training programs and raise awareness for agritech startup development.

- Provide preferential loans and establish patient investment funds for agritech.

- Support collaboration between agritech startups and agribusinesses.

- Streamline administrative procedures for medical treatments.

- Encourage collaboration between healthtech startups and medical professionals.

The Technology Startups Ecosystem

The term "startup" does not define a type of business, but rather, its stage of development. A startup is a collection of resources (including human resources, money, and time) that are combined into a new business model to quickly build it to a larger scale and replicate it in different markets.

The hallmark of a startup is its creativity, which distinguishes it from a traditional enterprise. A startup begins with a new idea or breakthrough and is often associated with a new or advanced technology. At the same time, a startup is also associated with the ability to commercialize and expand the market. That is, a startup must be able to put the idea into practice, attract customers, and have potential for rapid growth. Creativity helps a startup create new value and establish a new business model for the economy.

Although startups have the potential to contribute greatly to economic development, they face major obstacles to growth. The primary obstacle is the lack of capital, which is often needed in large quantities due to requirements associated with new technologies. In addition, startups often face major risks due to the novelty and creativity of their endeavor. Turning an idea into commercial reality requires a high level of intelligence and effort. In this context, startups are not able to develop on their own—they need an ecosystem to support them.

The concept of a startup ecosystem was first raised in the United States (US) in the late 1950s with the development of the Santa Clara Valley (the precursor to Silicon Valley), before being adopted globally in the 21st century. The Organisation for Economic Co-operation and Development defines the startup ecosystem as

> "a combination of formal and informal links between startups (potential or present), startup organizations, enterprises (companies, venture capitalists, angel investors, banking systems), related agencies (universities, state agencies, public investment funds), and relevant factors (the rate of enterprise establishment, the number of enterprises with good growth rate, the number of entrepreneurs), which have a direct impact on the local startup environment" (OECD 2015).

At the national level, a startup ecosystem can be described as the way a city, country, or territory supports the establishment of a set of entities—companies, venture capitalists, investment angels, banks, universities, government agencies, financial institutions—that are interconnected to foster technology-based entrepreneurship.

According to the World Economic Forum (WEF 2014), a startup ecosystem consists of the following nine components: (i) government policies; (ii) legal framework and infrastructure; (iii) capital and financial resources; (iv) culture; (v) consultants, advisors, and support systems; (vi) universities that act as catalysts; (vii) education and training; (viii) human resources; and (ix) domestic and international markets.

1.1. Overview

Viet Nam has a population of about 97.3 million—about 70% of which are internet users of which 98% are smartphone users. Internet usage is more than 10 times higher than in the last decade. Given these statistics, Viet Nam is considered a large technology market. However, the concept of a startup is still new. Startups first appeared in the country 15 years ago, when several companies were founded that used new technologies in e-commerce and online education. It was only in 2016, however, when all the ingredients of a tech startup ecosystem first became available in Viet Nam.

Viet Nam's 2017 Law on Small and Medium Enterprises (Government of Viet Nam 2017) defines a startup as "a starting process based on the creation or application of new technologies and management solutions to improve productivity and quality." Startups enable the creation of value-added products that allow for rapid growth in the market. Similarly, the Ministry of Science and Technology (MOST) describes a startup company as an "innovative start-up business established to implement a business idea based on exploiting intellectual property, new technology or business model and has the potential to scale up rapidly" (Phan 2021). Project 844 supports startups with an operating period of no more than 5 years from the date of issuance of the first business registration, while support for startups under the Law on Small and Medium Enterprises is not subject to this time limit.

According to Amway (2018), Viet Nam is a global leader in entrepreneurship and has a positive attitude toward starting a business. Specifically, 95% have a positive attitude toward entrepreneurship and ownership.

There are no official figures on the number of startups, but according to Namjatturas (2018), there are currently about 3,000 startups, almost double the 1,800 in 2015. MOST statistics in 2018 also recorded about 600,000 companies in the country, including 3,000 innovative companies. In 2021, VN Express, an international Vietnamese online newspaper, estimates the total number of tech startups to be about 3,800, with four unicorns—VNG, VnPay, SkyMavis, and MoMo (Thong 2021).

According to the Vietnam Chamber of Commerce and Industry (VCCI), the number of business startups per capita in Viet Nam is higher than in other countries such as the People's Republic of China (2,300), India (7,500), and Indonesia (2,100). It is widely recognized that such rapid growth is the result of a large team of talented managers, founders, product developers, and engineers, including significant contributions from overseas Vietnamese who have returned home to create startups.

A comparison between companies from different sectors highlights the outstanding features of information technology (IT) companies that influence the development trend in the context of the Fourth Industrial Revolution: (i) IT-based startups do not require much initial capital; (ii) they rely heavily on new ideas and highly innovative, fast-growing strategies; (iii) they are able to easily connect globally through technology and make their creative ideas accessible worldwide and vice versa; and (iv) companies easily learn from successful international models.

Viet Nam's startup community has achieved the success of first-generation companies formed in the early 2000s such as Vinagames and VC Corporation (Vatgia), and second-generation innovative startups formed around 2010. In 2016–2022, third-generation companies have emerged in the fields of education technology, agriculture, health services, financial technology, and e-commerce. Most third-generation startups are less than 2 years old, small, and not yet able to grow. These startups have investment capital of less than $10 million, which pales in comparison to the capital-raising activities of startups in other countries in the same sectors.

For Viet Nam startups, 2016 was a remarkable year. Seven funding transactions exceeded $10 million, and the total value of investments increased significantly compared with the previous years. In total, funding transactions for startups reached $205 million, 46% more than in 2015. In 2021, there were 17 deals whose total deal amount exceeded $10 million, an increase of 255% over 2020 (NIC 2021).

Figure 1: Total Number of $10 Million or Smaller Deals and $10 Million+ Deals, 2013–2021

(%)

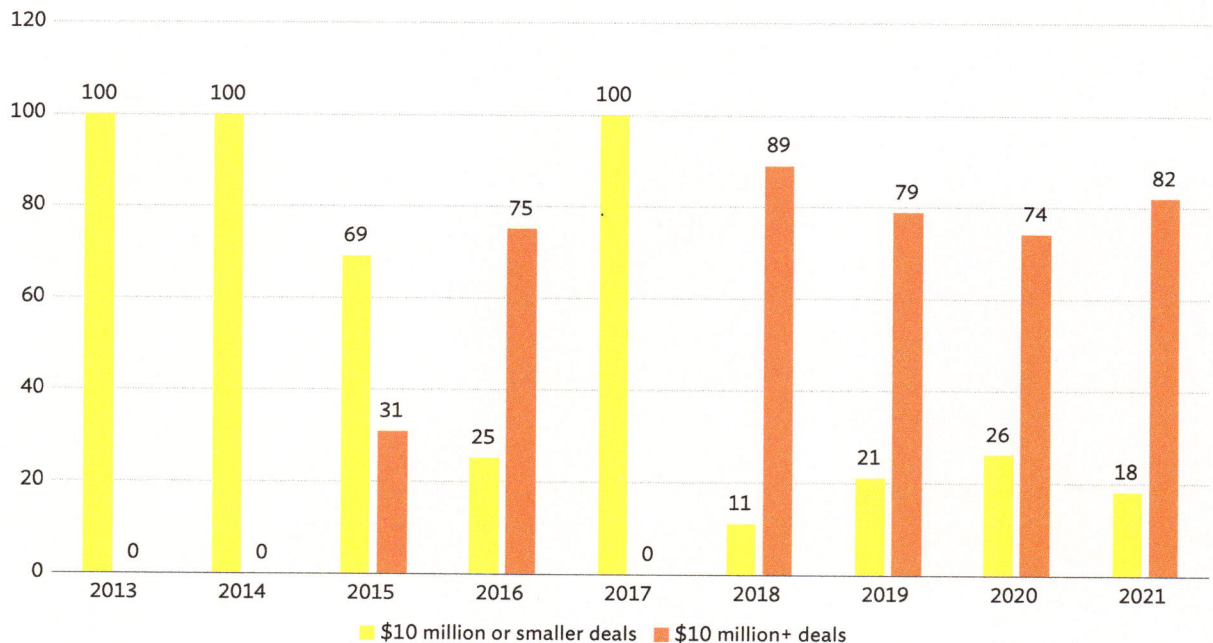

Source: NIC (2021).

Since 2018, startups in Viet Nam have become increasingly active. Viet Nam attracted $443 million in investments in innovative startups in 2018, about 10 times more than in 2017 ($48 million) (DO Ventures 2020). In 2019, the figure doubled to $874 million, an impressive increase that shows the strong growth and dynamism of the ecosystem. Since 2011, when VCCI began collecting investment-related data for tech startups, the number of investment deals has fluctuated. From 2017 to before the coronarivus disease (COVID-19), investment deals in Viet Nam have shown an increasing trend decreasing in 2020 due to the COVID-19 pandemic, before rallying back again in 2021. Most typical venture capital deals are with startups that have demonstrated initial success and have gained a foothold in the market. However, 2021 showed a record increase in funding in edtech (from $8 million in 2020 to $55 million in 2021) and healthtech (from $3 million in 2020 to $37 million in 2021), boosted by the COVID-19 pandemic (NIC 2021) (Figure 2).

During the government-backed Vietnam Venture Summit 2019 attended by many foreign investors, 18 foreign investment funds pledged to invest $425 million in Vietnamese startups over the next 3 years. For example, the Republic of Korea's

Figure 2: Recorded Deals and Capital Invested in Tech Startups, 2013–2021

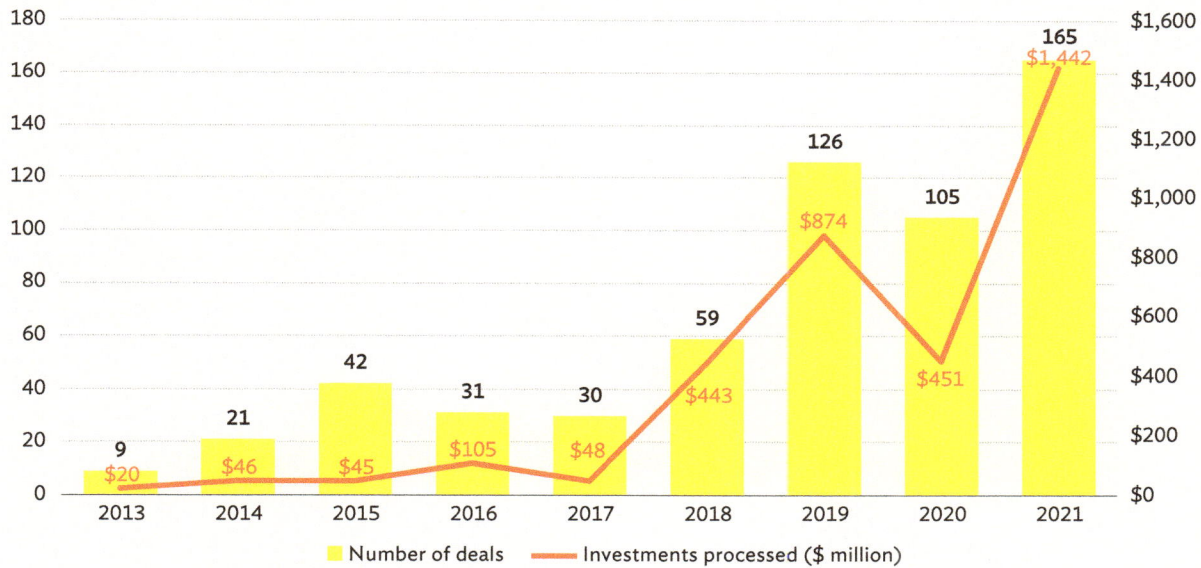

Source: NIC (2021).

investment fund, DT&I, invested $1.4 million in Propzy. VinaCap Fund has signed a strategic cooperation agreement with two funds from the Republic of Korea to invest $100 million in Vietnamese startups within 3 years as well. However, nationwide lockdowns due to the COVID-19 pandemic have reduced venture capital activity in the first half of 2020. Investments in Vietnamese tech startups rallied back in 2021, reaching a record level of $1.4 billion from 165 deals (NIC 2021). According to Nguyen (2022), in 2021 the top five sectors that received largest funding volumes are: fintech (26.6% of total), e-commerce (20.3% of total), edtech (17.2% of total), healthtech (7.8% of total), and software as a service (6.3% of total). These correspond to the top five leading startup areas for the whole economy.

1.2   Characteristics

The national startup ecosystem is focused on developing vibrant, fast-growing companies based on the use of intellectual property, technology, and new business models. The legal framework for investing in startups will be critical to building creative startup communities, networks of investors, and startup advisors, and in promoting interaction among key stakeholders. In addition, the government is gradually completing several policies to encourage investments in Vietnamese startups, demonstrating its commitment to improving the business environment and promoting entrepreneurship (Figure 3). The latest Global Entrepreneurship Monitor (GEM) special report on Viet Nam showed that 25% of the 2,118 Vietnamese respondents have entrepreneurial intentions, with 62.1% of the surveyed adults stating that becoming an entrepreneur is a desirable career of choice (GEM 2018).[1]

Figure 3: The Startup Ecosystem in Viet Nam

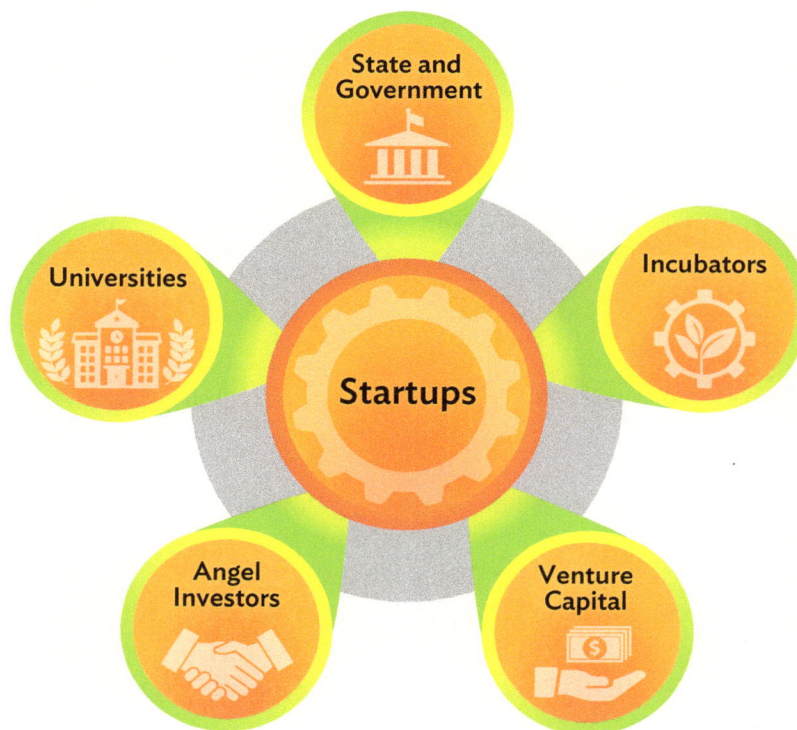

Source: Authors.

In terms of human resources, startup entrepreneurs have started to build closer ties and cooperation with local and foreign experts and Vietnamese experts abroad. For example, the Vietnam Innovation Network Initiative currently connects more than 100 scientists and technology experts. In addition, the Vietnam Mentors Initiative (VMI), a project to support startups, serves as a digital meeting place for startup advisors and consulting service providers (Van Roosmalen and Totten 2019). In 2018, VMI had 234 startup groups using startup consulting

---

[1] The latest GEM 2017/2018 special report for Viet Nam is based on a survey of 2,118 Vietnamese individuals and 36 experts. Founded by Babson College, GEM collects primary data through a population survey of randomly selected adults aged 18–64 in each economy. The most recent 2021/2022 Global GEM report covered 150,000 individuals in 50 countries.

services, 190 startup mentors, and 292 trained mentors. These are promising numbers for the development of human resources to support startups.

In terms of financial support, private companies have created several investment funds in line with the new legal framework introduced by the government, such as Startup Viet Partners, which raised $5 million in 2018, and SoftBank's $100 billion venture capital fund, which focuses on funding tech startups and innovative businesses (Viet Nam News 2019).[2] Large corporations have also increased their interest in local startups: Vingroup established VinGroup Ventures (with $300 million in investment capital) (Das 2019); and VinaCapital Ventures also established a $100 million venture capital fund to invest exclusively in tech startups (VinaCapital Ventures 2019; Vietnam Insider 2020).

The international community has also shown interest in Viet Nam's tech scene. Unitus Impact Fund is a VC firm that has opened an office in Ha Noi in addition to its offices in Bangalore, Jakarta, and San Francisco ($45 million in capital). Alpha Startups, owned by Malaysian company 1337 Ventures, has set up an accelerator and coworking space in Viet Nam, giving $5,000 each year to the top three teams in their cohort. The Finnish group, Tekes, in partnership with MOST, has launched the Tekes Innovation Fund to provide advice and product development support to startups in the areas of health and food safety, waste treatment, and education (Vietnam Advisors n.d.).

While local startups are actively building relationships with international partners, many lessons can be learned from the experiences of many government programs in countries such as Israel, Finland, the US, and Singapore. For example, MOST partnered with the US Embassy for the 2016 Innovation Roadshow. The American delegation consisted of multinational companies, financial investors, and successful early-stage US startups. They met with Vietnamese government officials, business leaders, venture capitalists, students, and startups (US Department of State 2016). Local businesses benefited from interaction with advanced startup models, which provided opportunities to connect with business partners and potential investors around the world.

In summary, while Viet Nam's startup ecosystem is still in the early stages of developing a culture of entrepreneurship, startup density, and supporting organizations, it is more advanced in terms of state policies, legal environment, as well as human resources for entrepreneurship. Even though the startup ecosystem is still young, it has already shown great potential for development. Every component of the ecosystem (from startups and investors to universities, incubators, and others) has made great progress since the early 2000s. Synchronous coordination among ministries,

---

[2]   See Crunchbase. Startup Viet Partners.

branches, and localities in formulating and promulgating supporting policies for startup activities has strengthened creative innovation in Viet Nam, thereby promoting the development of startups in both quantity and quality.

1.3 Startup Ecosystem

1.3.1 State and Government

The government pays close attention to the development of startups, considering them as one of the pillars for sustainable economic growth and improving competitiveness in the context of globalization and the Fourth Industrial Revolution. In recent years, the government has aligned, guided, and implemented many laws, regulations, and programs to support startups. In 2016, the Prime Minister approved the Supporting the National Innovative Startup Ecosystem to 2025 project (also known as Project 844) (Quynh 2018). This is the government's most direct intervention to create a favorable environment for startups in Viet Nam. With the mandate to implement Project 844, MOST is committed to achieving certain goals (Table 1).

Table 1: Project 844 Targets for the Development of Startups in Viet Nam

Type	Target
Legislation	Improve the legal system to support innovative entrepreneurship
Information	Set up the national portal "Start-Innovation"
Number of projects supported	Phase 1: 800 projects Phase 2 (by 2025): 2,000 projects
Number of businesses supported	Phase 1: 200 enterprises, of which 50 have successfully raised D1,000 billion in capital Phase 2 (by 2025): 600 enterprises, of which 100 have successfully raised D2,000 billion in capital

Source: Summary of interview with Project 844 officials, October 2019.

Startup programs and Project 844 are implemented at the city and provincial levels in large cities such as Ha Noi, Da Nang, Ho Chi Minh City (HCMC), and Can Tho, and in smaller cities such as Vinh Long, Ben Tre, Quang Binh, and Thai Nguyen. It is worth noting here that local governments also have their own startup/SME support programs (e.g., Saigon Innovation Hub or SiHub in HCMC has provided significant funding and other support to startups). Central and local policy documents have been developed that, while not mandatory, provide the legal basis to motivate local authorities to implement activities to support startups. In addition, the government approved two other projects in 2017: Supporting Students in Starting a Business Up to 2025 (known as Project 1665), implemented by the Ministry of Education and Training, and Supporting Women's Startups in the Period of 2017–2025 (Viet Nam News 2018; VietnamPlus 2017).

The Law on Small and Medium Enterprise Support (SME Law), which was adopted in 2017 and took effect in January 2018, officially establishes the legal status of startups. The main legal document is a decree issued on 11 March 2018 (Decree No. 38/2018/ND-CP), which provides detailed guidelines for investing in innovative startup SMEs, especially for the formation and operation of creative startup funds, providing an important legal basis for establishing organizations to support Vietnamese startups. The decree also sets the basic premise for the development of a capital market for innovative startups.

The government's plans place responsibility on ministries and relevant agencies to build an ecosystem conducive to the development of startups. The types of support for startups include the following:

- Support in application and **technology transfer;** the use of equipment and facilities at the technical facility; participation in incubation facilities and general working areas; and guidance in testing and perfecting new products, services, and new business models;

- Support in intensive **training** in construction and product development; attracting investment; intellectual property consulting; conducting procedures on standards, technical regulations, measurements, and quality;

- Support in **information,** communication, trade promotion, connecting networks of creative startups, and attracting investment from innovative startup investment funds;

- Support in the **commercialization** of results of scientific research and technological development, exploitation, and development of intellectual property; and

- Support in **investing** in startups and providing interest compensation through credit institutions (subject to the government's decision from time to time).

In addition, since startups are under the purview of small and medium-sized enterprises (SMEs), the government may provide them with the general support provided to SMEs, including the following:

- Support in accessing **credit** and the credit guarantee fund for SMEs;

- Support in **tax** and accounting matters and in obtaining information, advice, and legal assistance;

- **Technology** support and use of incubation facilities, technical facilities, and shared working areas; and

- Support in **human resources** development, acquisition of production facilities, and market expansion.

The government has also introduced standard programs to support startup innovation, such as (i) a project to promote innovation and creativity through research, science, and technology; (ii) a project to formulate policies for the renovation and development of business incubators; and (iii) the hosting of national and provincial events such as TechFest, Demoday, HatchFair, Venture Cup, Startup Weekend, and Startup Fair Danang.

The annual TechFest program has a broad national scope. In 2017, the program's theme was Connecting the Startup Ecosystem, which brought together relevant ministries and branches of Project 844, as well as startups and partner startup communities in Viet Nam and the region. TechFest 2017 included specialized areas in education and training, agriculture, medicine, travel, transportation, e-commerce, fintech, and games, entertainment, and media. TechFest 2018 launched the Service Constellation Initiative to Connect Vietnamese Startup Support Services platform, a hub that provides high-quality consulting services for innovative startups at preferential prices, with the participation of several reputable network partners such as Deloitte, OSAM, Duane Morris Vietnam, and Phusjion Group. TechFest 2019 aimed to link the fields of development and entrepreneurship in the four major economic regions. Techfest 2020 pushed through with the theme "Adapt-Transform-Breakthrough," while Techfest 2021 was themed "Embracing Innovation, Reshaping the Future." The most recent Techfest 2022 that was held in March continued to celebrate innovation in the spirit of previous techfests. It featured live and online activities and 2D and 3D virtual exhibitions.

In addition, several institutions have been established nationwide, to provide consulting and support services, and promote startups innovation. In 2015, MOST set up the National Technology Innovation Fund to attract knowledge, organizations, individuals, entrepreneurs, and scientists to contribute to economic development and participate in the creation of startups. It has an investment capacity of $47 million. Vietnam National University has launched the Youth Startup Program for 2016–2021 to promote youth participation in startup creation. At the launch, $22,500 was distributed to the 10 most promising startups. Both organizations are backed by the government.

1.3.2 Angel Investors

The number of angel investors in Viet Nam is still small, but it is increasing. These include local and foreign early-stage angel investors with small investments (several tens of thousands of dollars) and smaller angel investors who are often relatives and friends. Some active larger angel investors in Viet Nam include

- Vietnam Silicon Valley Accelerator, an accelerator and fund supported by MOST;[3]

- CLAS Expara Vietnam Accelerator (CEVA), an accelerator and fund initiated by Microsoft Vietnam for startup and early-stage (pre-seed) companies;[4] and

- Vietnam Startup Acceleration Fund (VIISA), which is both an early-stage investment fund and accelerator with two main investors, FPT and Dragon Capital Group.[5]

Most of these large angel investors are successful first-generation startup entrepreneurs who have invested in the next generation.

1.3.3 Venture Capital

In 2019, there were about 40 active venture capital funds operating in Viet Nam, up from 10 in 2015. Some well-known funds are 500 Startups, Dragon Capital, Mekong Capital, IDG Ventures, CyberAgent Ventures, Captii Ventures, Gobi Partners, and Zone Startup Ventures. In addition, recognizing their need to adapt to technology innovation to remain competitive, many Vietnamese corporations and large companies have established investment funds such as VinaCapital Ventures FPT Ventures, Viettel Venture, CMC Innovation Funds, and Accelerated Funds. There are also regional VCs that are active in Viet Nam, in particular, there is strong appetite from Korean VCs. A few private local VCs were launched in 2020-2021. In 2022, there are no specific regulations for governing the formation and development of venture capital funds in the country.

The number of investment transactions in Viet Nam is small but increasing. However, investment capital for innovative startups is still relatively modest compared with the region and globally. In 2017, the number of deals with investments of more than $10 million is very low, while most are under $1 million. There are also very few

---

[3] Vietnam Silicon Valley is a leading accelerator that invests in world-class early-stage startups.
[4] CLAS Expara Vietnam Accelerator.
[5] Vietnam Innovative Startups Accelerator (VIISA).

Figure 4: Investment Deals by Type, 2021
(%)

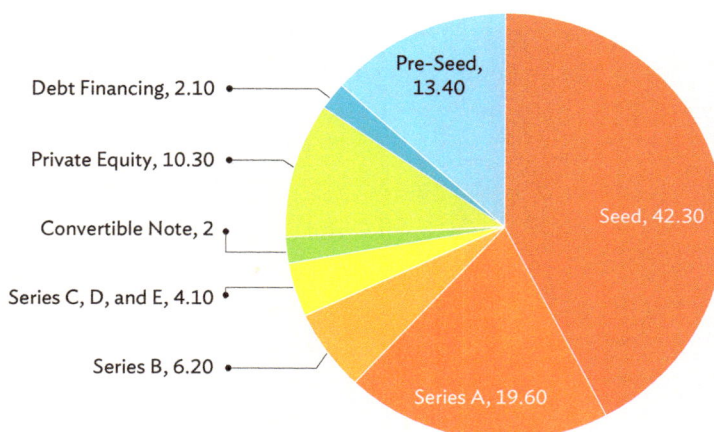

Pre-Seed, 13.40

Debt Financing, 2.10

Private Equity, 10.30

Convertible Note, 2

Series C, D, and E, 4.10

Series B, 6.20

Seed, 42.30

Series A, 19.60

Source: Nguyen (2022).

mergers and acquisitions (Figure 4). The total VC funding volume reached a record high of $1.4 billion, almost double the prior record of $874 million achieved in 2019 (NIC 2021). Seed (42.3%), Series A (19.6%), and pre-seed (13.4%) are the most popular deal types. Higher value deals in Series C, D, and E comprise only 4.1% of total deals but are expected to increase in the coming years.

In 2019, Viet Nam witnessed a leap in its venture capital market. While the country previously attracted about 3% of all venture capital in Southeast Asia, it now attracts 19%, rising from fifth to third place in the region, behind only Indonesia and Singapore. Vietnamese investors have also increasingly made their presence felt in the national startup ecosystem, accounting for 36% of total deals in the first half of 2019. In 2021, the top investors in the tech startup scene of Viet Nam are Singapore, Viet Nam (local VCs and angel investors), and the US (NIC 2021). But while the capital market for startups in Viet Nam has become more diverse, the level of investment is still low and does not meet the needs of the startup community.

1.3.4 Incubators and Accelerators

A startup incubator is a company or organization that takes care of startups by providing them with services so that the startup can overcome the difficulties in the early stages. Incubators have been established in most provinces to support the emerging startup community.

In 2018, the number of organizations supporting startups, including university-based incubators as well as foreign and local incubators and accelerators, that are primarily public sector funded was about 50—almost 50% more than in 2017. In 2020, this exploded to 43 incubators established by universities, institutes, and colleges, 38 incubators and 23 accelerators owned by the private sector mostly located in Ha Noi and HCMC (Nguyen and Le 2020). Major incubators include Hoa Lac Computerized Numerical Control (CNC) Incubator, Ho Chi Minh City CNC Business Incubator, Danang Business Incubator (called DNES), Center for Youth Economic Information Consultancy's Business Startup Support Center (BSSC), and Hanoi Incubator for Information Technology Innovation.

Some accelerators—such as the Vietnam Silicon Valley Accelerator, which is supported by MOST, and VIISA—focus on early-stage startups and provide them with funding. The number of coworking spaces for startups has increased to 70, most of them in Ha Noi and HCMC, and they meet the startups' needs for material and technical facilities as well as training and networking.

The National Innovation Center (NIC) under the Ministry of Planning and Investment (MPI) was established in 2019 and is now in operation. The center is an important component of the National Network of Startup Entrepreneurs and functions as a high-tech business incubator that promotes and supports individuals, startup organizations, and investors; improves the capacity to receive and apply (new) technologies; and enhances the capacity for research, development, and innovation of startups in the context of the Fourth Industrial Revolution.

Other activities that connect the domestic and international startup ecosystem and promote culture of entrepreneurship include events, entrepreneurship programs, and international cooperation in the field of respective startups. Some of these events are (i) Startup Wheel 2016 and Startup Ideas 2016, organized by BSSC and Young Entrepreneurs Association of HCMC; (ii) National Entrepreneurship Program organized by Vietnam Business Forum newspaper together with Vietnam Chamber of Commerce and Industry (VCCI); and (iii) Startup Festival 2016 "Pioneer Aspiration" organized by the Youth Department of Vietnam Television, Vietnam Climate Innovation Center (VCIC), Topica Founder Institute Education Complex, and Blue Bird Joint Stock Company.

Although incubators and accelerators have been active in the last 10 years, they are still a relatively new concept in Viet Nam. Despite the support mechanisms and policies, incubators, especially public incubators, have not been able to fully meet the development needs of these startups. Many incubators have been successful, but the strong development of the startup community has revealed their limited capacity. For example, most of these incubators are young, having been established only between 2015 and 2018, and they are located in only four major cities (Ha Noi, HCMC, and Da Nang).

Some provincial accelerators are highly specialized and support only certain startups, such as those led by women or ethnic minorities, or those that promise social impact (Nguyen and Le 2020).

1.3.5 Universities and Research Institutes

Universities play an important role throughout the startup process, from idea formation to product development to the growth phases of startups. In the first phase, lecturers and support units provide inspirational guidance and are the source of information, examples, and lessons learned. They highlight successes and support continuous team development and improvement by promoting interdisciplinary collaboration among students. If the businesses already have products and services, the university provides the basic business-related knowledge related to laws, taxes, and accounting. Then, in the third phase, when the ecosystem has many good startups, the university provides skilled human resources with exemplary thinking skills and facilitation experience for sustainable business growth.

Universities and research institutes in Viet Nam, in cooperation with provincial authorities and central ministries, have set up units to support startups within universities and organize startup events for students to promote the culture of entrepreneurship among young people. These events include the Startup Student Ideas Contest of the Vietnam Student Association; Starting a Business with Kawai (Foreign Trade University); I-startup (National Economics University); and the Agricultural Startup Contest (Vietnam National University of Agriculture), all of which are designed to promote entrepreneurship among young people.

In addition to several advisory networks, university-based coaches, startup advisors, and mentors also provide advice and support to young startup entrepreneurs. For example, the Vietnam Mentors Initiative (VMI) was established in 2016 to connect mentors with other mentors. Many startup organizations have joined VMI, whose vision is to become a coalition of startup advisors with the most widespread and effective network in Viet Nam.

To help them build and commercialize their ideas and business plans, students also have access to many entrepreneurship competitions. Typically, students are encouraged to start their own businesses, preferably a startup, beginning in their second year.

Another positive factor is that Viet Nam ranks well in the Global Competitiveness Index and the Global Innovation Index. This is the result of reforms in research funding, university autonomy, and the commercialization of research results, including the use of technology transfer centers.

Survey of Agritech and Healthtech Startups

2.1   Viet Nam's Agriculture and Health Sectors

In 2018, the export value of Viet Nam's agricultural, forestry, and fishery products reached over $40 billion. In 2022, export turnover reached a record high of $48.6 billion, an increase of nearly 15% compared with 2020. The agriculture sector accounts for 15% of gross domestic product (GDP).

However, agricultural production in Viet Nam remains fragmented. Vietnamese products are of low quality and value added, and the country's ability to respond to disasters triggered by natural hazards is still inadequate. In general, the agricultural industry has not yet fully applied high technology and still uses manual techniques. Investment in research is insufficient, and the country's agricultural products do not have much of a competitive advantage in international markets. Domestically, the sector has not built a network of consumer markets that can adequately regulate supply and demand. The linkage between businesses is low, and most agricultural producers produce low-value agricultural products.

However, consumers' awareness of food quality is increasing, and they are increasingly demanding to know product origin and other related information. However, the system of product quality standards is not extensive, making it difficult for farmers to ensure credibility and product quality commensurate with price. The sector relies on the use of pesticides, herbicides, and chemical fertilizers, and organic agriculture has only recently become popular.

The government has retained a special interest in the agriculture sector and has issued many policies to support agricultural startups and businesses that invest in agriculture and rural areas, including those that promote the application of high technology throughout the agricultural production chain to create added value. However, only a modest number of businesses have access to those policies. As of 2017, the government has provided more support to the sector, which could lead to more investment. Many countries provide official development assistance loans for agriculture and have increased their cooperation with Vietnamese partners to build value chains.

Agriculture should be a promising sector for tech startups. Despite the goal of becoming an industrialized and modern country, much of Viet Nam's economy remains agricultural. Much of the population is young and moving toward an open economy, with agricultural investment coming from overseas. The Fourth Industrial Revolution and the increasing demand for organic and safe agricultural products have created new opportunities and markets favorable to startups. In addition, growing demand for safe and nutritious fresh or processed food, including delivery services and digital payment platforms are new market trends that agritech startups could capture.

With a population of more than 90 million and only slightly more than 1,000 hospitals, Viet Nam's medical development needs are enormous. According to the latest data from World Health Organization's Global Health Expenditure database, the country's health expenditure was 5.25% of GDP in 2019, down from 5.5% in 2017.[6] Per capita health-care expenditure in current US dollars was $558.87 in 2019, three times higher than 10 years earlier.[7]

Viet Nam has enormous potential for the use of medical information technology. There have been few startups in the health sector. In addition, the high proportion of people using smartphones and increasingly more having internet access can provide a potentially large market for digital-based solutions for health-care services.

The healthtech sector in Viet Nam is focused on providing services. Health-care startups often deal with noncommunicable diseases. Health-care apps provide services such as a search function for finding information, signing up for a doctor's visit, providing health histories, connecting patients with doctors and pharmacists, and follow-up care. These services come with the latest medical technology that effectively connects medical doctors and citizens and reduces intermediate costs.

Another trend among healthtech startups is the focus on preventive measures such as health monitoring and the creation of data connections and links between health-care facilities, which essentially help to create an integrated health-care and consultation system. Currently, health information technology (IT) systems do not share patient data with each other for information security and privacy of health records. Hospitals also use different solutions and software, so data connection and sharing is relatively challenging. There is great potential for health-care startups in the country to meet the need for innovation in services such as data and information sharing among health-care IT systems and provide comprehensive health-care solutions for the community.

---

6   World Health Organization. Global Health Expenditure Database (accessed February 2017).
7   The World Bank. Current Health Expenditure (% of GDP) – Vietnam (accessed 2018).

2.2 Interviews with Startups, Ministries, and Other Stakeholders

In establishing the list of startups to be interviewed, the research team did not find any greentech and edtech startups in the list of startups to interview. The few greentech and edtech startups that were approached were not willing to be interviewed. Hence, the team presented results for a sample consisting of only agritech and healthtech startups. This sample was sent to the Asian Development Bank for confirmation.

The study team conducted a survey of 11 Vietnamese startups, five of which were from the agritech sector and six from the healthtech sector (see Appendix 2 for a description of these startups including the type of services they provide).[8] Where possible, the startups' founders were interviewed, or in their absence, the cofounder or next in a leadership role. All startups had been established for less than 5 years at the time of the interview.

Before the face-to-face interview, the study team sent the startups a simple questionnaire to complete. The questionnaire focused on the startups' growth constraints. This was followed by an in-depth interview at the startups' offices, which lasted between 30 minutes and 1 hour, depending on the founder's availability. Interviews were conducted at the startups' offices in Ha Noi or HCMC in Vietnamese, translated into English as needed, and transcribed with pen and paper; no recording device was used.

Interviews usually begin with the consultant giving a brief introduction to the research project, followed by the startup founder introducing the tech startup. Discussions were informal and aimed to answer the question, "What factors helped, and what factors constrained the growth of your startup during the building and scaling phases?" These factors relate to the elements of the startup ecosystem shown in Figure 3. Thereafter, the interviewer thanked the respondents for their time and assured them of the confidentiality of their responses. The results of the survey and discussions are presented in the next sections.

In-depth interviews with other stakeholders were conducted in person. Among the government agencies visited were the National Agency for Technology Entrepreneurship and Commercialization Development (NATEC), the State Agency for Technology Innovation (SATI), and Project 844. Interviews were also conducted at the Bach Khoa University Innovation Center, the country's premier university for science and technology, which hosts an incubation center for its students.

---

[8] The survey was conducted in September and October 2019. The prepandemic scenario is assumed to reflect business-as-usual case for the startups.

2.3 General Characteristics of Startups

All of the interviewed agritech and healthtech startups are small—agritech startups have an average of 20 employees, while the healthtech startups have 23 workers. The smallest startup had 10 workers, while the largest had 39 workers including the founder(s), regular staff, and temporary contractual staff (Figure 5).

About 40% of the staff in agritech startups work in the office (design, accounting, marketing, sales), while the rest are assigned in the field (for testing, laboratory, production, as well as delivery, product installation, and repairs). On the other hand, 25% of human resources in healthtech startups work in the office, while 75% work in the field (delivery, installation, troubleshooting). Healthtech services are generally delivered through an online platform, while agricultural products require on-farm or in-person services. Generally, administrative staff (for office or accounting) is kept to a minimum and often outsourced to freelancers on a monthly basis to minimize expenses.

In both sectors, most founders have bachelor's degrees in science or IT rather than in humanities and social sciences. The founders are young (all are under 30 years old) and usually spend most of their time on technical solutions to develop the product.[9] Many founders had their technical ideas while studying at universities or technical

Figure 5: Startup Size Based on Number of Workers

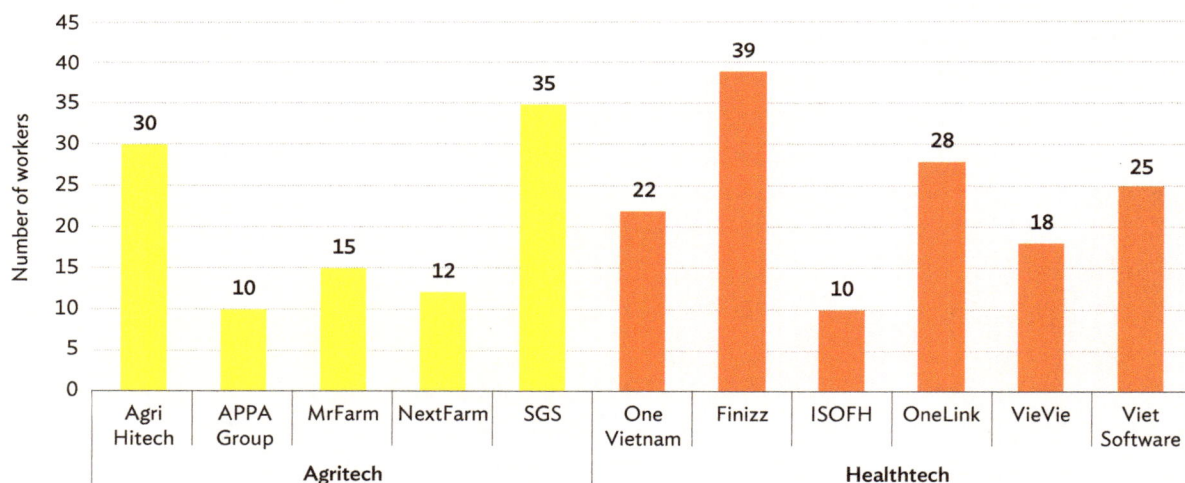

Source: Asian Development Bank.

---

[9] This study recognizes that many successful startups also have more mature founders (>40 years old) that have prior industry experience.

schools, while marketing ideas came much later. Respondents indicated that they need to spend more time on marketing and solving financial problems in order for business to develop and survive.

Recruiting and maintaining adequate human resources is a common problem for all startups, as they cannot pay high salaries and are not willing to give their employees equity in the company. This is an even more serious problem in the healthtech sector, where some medical knowledge—a hard-to-find qualification—is required. As a result, salaries in healthtech startups tend to be higher than in agritech startups. Nevertheless, healthtech startups are more prone to staff turnover, one of the main reasons cited for the failure of many startups in this sector.

However, these startups share a common strength: their founders are under 30, tech savvy, and passionate about commercializing new technologies. Many of them graduated from a top national university or a highly ranked technical university, which gave them access to modern technology, especially IT.

Some startups have had the experience of participating in international competitions or exhibitions that have received positive assessments, which gives founders the confidence to pursue their goal of capitalizing on their product. Some products have also attracted the attention of venture capital firms or large companies.

However, market penetration and financial sustainability pose a problem for startups. All of the interviewed startups said they considered their performance to be lower than their potential. Most startups also indicated that their lower-than-expected turnover rate makes it difficult to cover operating expenses. A common problem is often a lack of knowledge and effort about the business aspects of the startup.

None of the startups interviewed focused on branding and intellectual property rights (IPR) issues. Although all are aware of the importance of branding, IPR was not their priority at the time of the interview. They perceive the application process of IPR as complex, expensive, and time-consuming. All are aware of the weak implementation of IPR in Viet Nam.

2.4   Financing Needs

The critical factor for the survival and development of startups is capital. In addition to their small size, all interviewed startups have limited equity. Some startups generate modest revenues but do not have sufficient cash flow to cover the costs associated with expanding their market. Figure 6 shows how respondents rate their startups' current operating capital versus their capital needs.

Figure 6: Operating Capital and Capital Needs of Interviewed Tech Startups

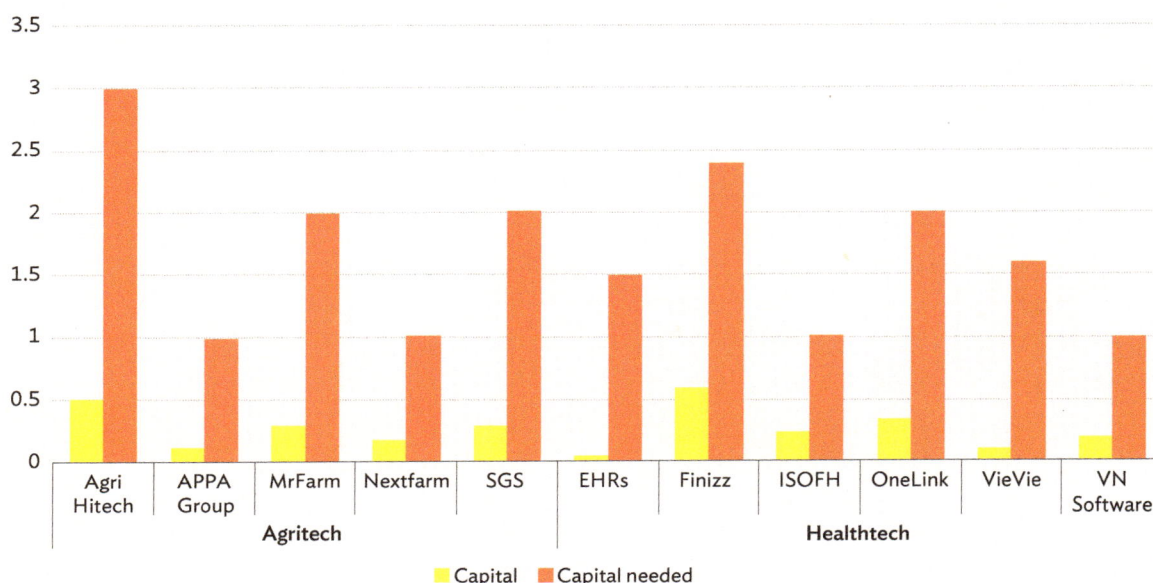

Source: Interviews by authors.

Startups require more capital than their current operating capital. They often use their own funds to raise initial capital or turn to their family and friends to obtain it. However, these are usually small amounts (less than $10,000) that do not get them very far.

After that, startups try to obtain funding from special funds offered by government agencies. Most of them are able to access government funding through government programs for SMEs and startups, but even these funds are not enough to scale the business. According to respondents, one reason for the low level of funding is the low level of trust that government funds have in Vietnamese enterprises. However, respondents from the ministries reported that startups quickly and easily promise certain actions or results if they choose to receive government funding, but fail to implement the required actions or take no responsibility for not doing so. Representatives of state agencies said that while investment funds are very willing to invest in startups, some of them require proof of reciprocal capital from the tech startup to ensure that the tech startup has the capacity and is able to take responsibility for the success of the investment. But in fact, the startups have very little of their own capital to return private investments. Some of the tech startups rely entirely on government investment to survive.

The third option for startups is to turn to banks. As part of the government's ongoing support for SMEs under the SME Law, which came into effect in 2018, some banks, independently or in cooperation with the government, are offering concessional rates to innovative SMEs, including tech startups. However, from the perspective of financial institutions, no distinction is made between SMEs and tech startups, as the loan terms are the same for both, such as interest rates, repayment rates, and loan terms. Tech startups that are in the development or testing phase are more likely to need the more patient types of capital or, in other words, capital investments that are not expected to yield immediate returns. They also have difficulty borrowing from banks, as they usually require collateral for larger loans and a feasible business plan and proof of the company's liquidity for business and commercial loans, which is not yet the case for tech startups.

The last funding option mentioned by startups is venture capital. However, venture capital funds are generally skeptical of startups and vice versa. According to the founders, VCs are willing to invest but require that startup profits are already high or scalability and potential to generate revenue is evident, a condition that is difficult for startups in the product development phase to meet. In addition, the founders shared that at least 50% of the startup's equity should be pledged to the VC firm in order to receive a substantial VC investment. The perception that control of the startup is lost after the acquisition of VC was a problem cited by the startups interviewed, although admittedly this experience could be unique because VCs typically ask for a minority stake. This perceived "loss of management control" makes many startups hesitant to receive venture capital. Conversely, however, government respondents said this perceived loss of control is unfounded. Instead, tech startups should focus on developing a proof of concept and, more importantly, a minimum viable product, as well as a clear and well-thought-out business plan to attract VC investment. Not all of the interviewed startups have both. Many startups are not able to develop a sustainable business model. Many do not have a long-term business plan and focus only on content for immediate needs. Few invest enough time and effort to develop a business plan and funding proposal or conduct a market analysis. As a result, startups are often surprised by market developments when revenues do not meet expectations. Another problem is the inability to manage cash flow. Many young entrepreneurs spend money quickly before they get good results. The main mistake startups make is that they are too optimistic about their revenues and profits. Weak financial management is not unique to startups, but is a common problem among small enterprises.

The initial findings of the interviews are consistent with the findings of the Global Entrepreneurship Monitor and the Vietnam Chamber of Commerce and Industry (VCCI) (GEM and VCCI 2018). This report stated that innovative businesses in Viet Nam (including tech startups) often face the following problems and constraints:

i.   **Capital constraints.** Startup businesses often begin with the limited capital of founders, as they are unable to borrow from banks or secure investment funding.

ii. **Limitations on skills.** Startups often lack technical and technological skills, enterprise management skills, product promotion, and marketing skills.

iii. **Limited ability to comply with administrative procedures.** Startup businesses often have very little experience in complying with administrative procedures related to market entry (e.g., business registration); intellectual property protection (e.g., patent registration); commercialization (product registration and meeting technical regulations); and business processes (accounting standards, invoices, tax declaration, securing tax incentives).

2.5   Barriers to the Growth of Agritech Startups

The questionnaire sent to agritech startups before the interviews gave them the chance to identify the opportunities and challenges they face as young businesses. Respondents agree that Viet Nam has always been an agricultural country, so the opportunities for entrepreneurship in the agriculture sector are clear. In fact, the agriculture sector has received more support from the government since 2017, which has attracted large enterprises to invest in high-tech agriculture. Viet Nam's large agriculture base shows that there are new opportunities for businesses at every node of the value chain. Currently, there is an increasing demand for "safe" agricultural products that are grown with minimal use of chemicals and whose origin is clear and traceable. Quality assurance is becoming an urgent need in the agricultural produce market, which emerging agritech startups can tap.

The sector also has its constraints. The three most commonly cited are (i) the weak physical and legal structure, (ii) the low connection of agriculture to innovation, and (iii) the low capital investment in agriculture.

The first constraint mentioned by respondents in the interviews is the weak physical and legal infrastructure. Viet Nam has an emerging digital infrastructure system with uneven connectivity quality. Cities have better digital connectivity than rural areas. This inequality contributes to weak information infrastructure connectivity nationwide, and cities and regions are not equally ready for digitization. For this reason, internet-based agritech startups have difficulty covering the rural areas where agriculture is most prevalent.

Many legal gaps also hinder the overall development process of tech startups, as identified by agritech and healthtech startups. These gaps are discussed in the following section of the report.

The second most common constraint is the lack of research in agriculture and technology that could serve as a theoretical basis for startups. Cutting-edge research in agriculture is underfunded, even though the sector accounts for 15% of GDP (as of 2018) and employs a significant share of the labor force. Despite the government's special interest in the agriculture sector and the consumer taste revolution in fresh produce, research and innovation in the sector could still be improved. The new knowledge created by scientific research in projects and universities could be brought closer to society and businesses so that these ideas can be used profitably. Respondents agree that this is one of the reasons why there are few successful agricultural startups in Viet Nam. Viet Nam agritech has focused on developing "rudimentary products and services" using simple technology and models, with very little investment. The marketing strategy is spontaneous and targets only end users. At the same time, many solutions could be offered along the production and distribution chain. The reason for this is the fact that few startup founders have a high level of ICT skills and still want to work in agriculture. The lack of skills also leads to a low level of innovation in this sector.

Some respondents mentioned that one way to overcome the low level of innovation is for tech startups in agriculture to join together to form an ecosystem or a tight commodity value chain where information can be shared. Although the agribusiness movement is very active and the Vietnam Farmers Union has 73,000 members, there is little interaction and active connection between members to support each other.[10] As such, agritech should be approached with broader solutions to solve issues not only at the farmer's level but also multi-sector issues related to information asymmetry, financing, and logistics.

Finally, the founders identified the weak confidence of foreign investors in agritech startups as a constant obstacle. Many foreign VCs are more interested in investing in startups in more lucrative sectors such as e-commerce, fintech, gaming, travel, or enterprise solutions. As a result, it is easier for agritech startups to approach local corporate VCs than foreign VCs. NextFarm, one of the respondents, stated that at the time of the survey it was operating with funds from corporate VC, which it acquired from one of the larger agriculture conglomerates in Viet Nam.

---

[10]   See Vietnam Farmer's Union.

Due to the cyclical and seasonal nature of agriculture, agricultural startups often require long-term investments, especially when it comes to testing technologies and prototypes, making it problematic to make immediate profits as in a traditional business. The survey results show that in Viet Nam, not many agritech startups have paid enough (or early enough) attention to these factors. Most startups need patient capital for their technology investments, but many have no experience in raising capital, while some do not even know how much they need. When asked how much capital their project needs, they are unable to provide an answer, which quite often causes them to miss out on business opportunities.

2.6   Barriers to the Growth of Healthtech Startups

With only about 1,000 hospitals across the country serving a large population, the need for medical development and health care in Viet Nam is high and increasing. The potential for the use of ICT in the health sector is enormous. While other sectors are saturated, there are not enough companies in the health-care sector to provide the range of services needed, especially since the administrative requirements for health care are very high. Viet Nam has many gaps to fill in the digital health space for development startups, and the government strongly supports technologies aimed at improving the health-care system and its services.

Thus, Viet Nam's healthtech sector has immense potential but faces many challenges such as (i) the widespread use of traditional services, (ii) weak physical and legal infrastructure, (iii) the lack of medical expertise among founders and their dependence on medical service providers for quality of information, and (iv) the lack of financial skills.

The custom of using traditional medical services is still popular, especially in rural areas where many are hesitant to switch to modern medicine and are unaware of new forms of services such as online health care. Vietnamese people are not in the habit of getting proactive and regular medical checkups and tend to go straight to the hospital for treatment when symptoms first appear. It will take a long time for this habit to change, as well as the habit of using free (or government-subsidized) rather than paid (fee-for-service) consultation services. While health startups depend mainly on user fees, many Vietnamese are aware of and use medical consultation services that are subsidized by the government's national health system. Healthtech startups need patient capital funds to cover the costs of doctors and health professionals and the use of online platforms.

In addition, infrastructural and legal barriers are a major obstacle to the growth of the sector. The process of digitizing administrative and medical data in Viet Nam is slow, data protection for medical data is strict, and hospitals still keep written records. For healthtech startups to succeed, some kind of national digital health database must first be created, which does not yet exist in Viet Nam. In addition, there are virtually no regulations for online medical consultation services and mechanisms for connecting, sharing, using, and providing information between health sector units, and between the health sector and related departments.

Although the government is making efforts to use information technology for one-stop government management, the application has not been synchronized, especially in the health sector.

A common problem of startups in the health sector is the lack of medical expertise. Most founders come from the ICT sector and have a weak foundational knowledge in the field of health. To be able to offer innovative solutions in this area, knowledge of both ICT and health is required. The founders' lack of clinical and hospital-specific background has hindered the development and use of medical-based apps. In addition, startups cannot proactively control the quality of their services, but depend on medical service providers for technical information and basic data. In fact, although health apps provide great technical support to users, the actual service is still provided by the clinic/hospital. In general, customers of healthtech startups often prefer traditional methods of seeking medical services at clinics/hospitals, even if they have already had access to more efficient online health services from startups.

Medical startups often require powerful systems and big data centers to process the complex, large, and multidimensional metadata and statistical information. However, health startups often have limited capital to develop these large databases and have little capacity to borrow from banks or apply for investment funds. A tight budget and a lengthy development process also make it difficult for them to retain staff. Staff with health technology skills are scarce and often sought after by large companies and foreign investors offering higher salaries. The constant loss of staff forces health startups to train new replacements. They may also struggle to retain and develop their staff.

Table 2: Summary of Constraints on Growth of Healthtech Startups

Challenges
Widespread use of traditional services. Vietnamese people are not in the habit of proactively seeking regular medical checkups and go directly to the hospital for treatment only when health problems arise.
Weak physical and legal infrastructure. Weak physical and digital infrastructure hinders growth. Legal barriers are very high. The process of digitizing medical data is slow. There are no regulations for online consultation services.
Lack of medical expertise. Most startup founders have a technological background but little medical knowledge.
Limited budgets. Patient capital is needed to build databases and networks.

Source: Interviews by authors.

2.7 Policy Environment Issues

The questionnaire and in-depth interviews put special emphasis on the legal environment in which tech startups operate to reveal the constraints that can be addressed through policy reform. Tech startups identified the following policy constraints.

i. **Foreign ownership limits.** In June 2015, the government signed Decree No. 60/20015/ND-CP, which raised the cap on foreign ownership limits (FOLs) in Viet Nam's public companies from 49% to an unlimited level. However, FOLs for restricted sectors remain in place. The framework for the restricted sector has yet to be finalized (Chung et al. 2015).

ii. **Specific tax incentive schemes.** Existing tax incentive schemes to encourage investment in Vietnamese enterprises in Viet Nam, such as enterprise income tax (EIT), preferential rates, EIT exemption, and EIT reduction, are granted based on location, industry, and manufacturing size.[11] There is no EIT incentive for venture capital investments unless they are made in startups or sectors that fall under the qualifying categories.

iii. **Differential treatment of offshore and domestic VC funds.** Offshore funds are considered foreign investors in Viet Nam and are therefore subject to the normal restrictions accorded to foreign investors. These include (i) market access restrictions under local law or international treaties to which Viet Nam is a party; (ii) a longer investment approval process compared with domestic investors; and (iii) capital remittances in and

---

[11] For example, sectors qualified for incentives include high-tech industries, scientific research and development, software production, education and training, medical services, sports and cultural activities, and environmental activities. In terms of location, economic zones, high-tech zones, and poorer areas are eligible.

out of Viet Nam are subject to foreign exchange regulations, meaning the capital must be processed through a single capital account in Viet Nam. Domestic funds are not subject to such restrictions. Currently, there are about 40 venture capital funds investing in startups in Viet Nam, but most are foreign funds with representative offices. Because of the weak and nonpreferential legal corridor, foreign venture capitalists will not choose Viet Nam as their first destination, but instead go to other countries in Southeast Asia where investment conditions are easier.

iv. **Overlapping policies for SMEs and startups.** Policies to promote the development of innovative enterprises overlap with other policies, especially those for SMEs.

v. **Cumbersome procedures do not meet the specific needs of startups.** Applying for intellectual property rights (IPR) or copyright is time-consuming in Viet Nam; nevertheless, the issue of IPR protection is very important for startups. Currently, Viet Nam's intellectual property registration provides only weak protection. Even if intellectual property is already registered, in many cases counterfeiting or theft of information for commercial purposes still occurs, and the authorities rarely take positive action. Therefore, many businesses strive to create technological barriers to protect their innovations.

While some of the above factors have improved in recent years, the problems discussed show that components of the Vietnamese startup ecosystem still need improvement.

Analysis and Policy Recommendations

Viet Nam is increasingly becoming a hub for startups in Southeast Asia, competing with regional leaders in creating a favorable ecosystem for startups. The startup survey and in-depth stakeholder interviews show that the country could very well be the next People's Republic of China or India in terms of the number of startups that could achieve unicorn status if barriers to ecosystem growth are addressed. The most pressing issues identified by the research team are presented below.

Startups in agritech and healthtech have identified a number of sources for early-stage funding, but it appears that serious venture capital funding is elusive for reasons related to demand or supply. Many startups feel that current government policies on access to funding are focused on small volumes and commercialization of mature technologies rather than funding technological innovation. However, before technologies mature, startups need patient capital because it takes a long time to test their technological innovations. Access to credit is difficult because most young startups are micro in scale and have little capital and almost no collateral for bank loans. In addition, startups are also high risk, making it very difficult to access traditional capital mobilization channels through commercial banks. Startups also expressed the need for improved financial literacy, but current startup policies are not strongly focused on improving the financial or management skills of founders.

In the context of the weak VC pouring into startups such as agritech and healthtech, the overall legal framework for private sector support of startups may need to be revised. The results of the interviews showed that there are no specific regulations for the creation and development of venture capital funds. The issues discussed earlier, such as the differential treatment of foreign and domestic venture capital funds, are legal issues that need to be resolved. The lengthy investment approval process and the subjecting of foreign venture capital to foreign exchange regulations may also need to be reviewed. At the provincial and sectoral levels, the support structure for private sector involvement in startups is not developed and depends largely on initiatives at the national level. In general, many of the current regulations on administrative procedures for private sector participation at the provincial level may need to be revised with a view to facilitating businesses for tech startups. The government and the Ministry of Planning and Investment need to consider reforms to administrative procedures to eliminate unnecessary procedures and promote online administrative procedures to help businesses, including startups.

Legal provisions and policies supporting startups in Viet Nam may need to be updated and harmonized across agencies. Most legal provisions target SMEs, and few provisions apply to startups. While SMEs and startups may seem similar at the initial stage, there is a clear difference between the two, which should be filtered into distinct policies. With the exception of Project 844, which sets specific quantitative goals for startup development, other policies do not have binding responsibilities for startup development success. For example, other ministries, agencies, and organizations charged with formulating annual plans to support innovation for startups do not have specific activities to implement Project 844.[12] At the local level, although the implementation plan for Project 844 has been issued, actual implementation is still in its early stage, so results cannot be assessed. What is known is that support from government agencies lacks detailed and enforceable regulations, which is reflected in unclear implementation plans in some communities. Their plans simply repeat the content of Project 844 without developing specific local measures for startups.

The scale of support for startups appears to be limited, focusing only on business incubator support, credit guarantees, business management consulting policies, and improving production efficiency. University-led incubators are neither fully equipped nor sufficiently skilled to support founders. Tech startups are encouraged as research projects that eventually lose their momentum when the student leaves the university or the subject is completed for the semester. Many universities abide by their mission of training students so that they find jobs, but may instead want to train students to take risks, experience new ideas, and create jobs. These approaches go far beyond the structure of university research projects. Instead of theoretical teaching, schools may want to combine theory with experiential learning or the application of knowledge. This should help students understand the mindset of entrepreneurs and how to make the best decisions in a given context. In the context of creating an enabling ecosystem for tech startups, in addition to training students in science, technology, engineering, and mathematics (STEM), universities may need to modify their competency profile by becoming more involved in supporting startups by engaging with policy makers and businessmen in their activities. This cohesion primarily benefits the university by making the school an indispensable element in the national innovation system.

---

[12] The current tax policy supports businesses by geographic area, so only an enterprise or startup that qualifies is eligible for the incentives.

The analysis shows that there is a need to establish a mechanism that links the key players in the Vietnamese startup ecosystem so that policy reform and activities can be drawn and managed in a unified system. The startup ecosystem needs a focal point with integrated functions and tasks such as (i) providing relevant information on ecosystem actors and activities; (ii) providing opportunities and information on training and cooperation opportunities; and (iii) providing guidance and organizing unified actions. This mechanism will be responsible for monitoring and updating information on the components and actors of the national startup ecosystem network. It will also continue to develop the ecosystem, focusing on the development of the framework legislation for technology startups, especially in terms of venture capital investments, tax incentives, and access to financial instruments. At the moment, Project 844, although a smaller and time-bound program, seems to be a potential mechanism to anchor the different actors of the startup ecosystem. The development of this network could be included in the mandate of the office.

3.1 Recommendations for the Creation of an Enabling Startup Ecosystem

Continue to remove barriers to business and cumbersome administrative procedures. There is a need to review the documents implementing the Law on Support for Small and Medium Enterprises (SME Law), including the programs therein that overlap with the policies and programs for innovative and technology-based startups. It is important to evaluate the financial needs of tech startups and develop financial programs that can support their unique growth. One example is to clarify and revise tax incentives for investment in Vietnamese startups. Certain sectors may need to be opened up to preferential tax rates to encourage investment.

Focus policies and programs on encouraging innovation. The government could continue to support scientific research and strengthen information exchange and capacity building by establishing science and technology development funds and providing support and guidance to enterprises in accessing these funds. It could strengthen and continue its support for translating scientific and technological breakthroughs into commercial production. Such startup support programs could go beyond traditionally identified investment sectors and target emerging sectors such as processing and digital services.

Strengthen intellectual property rights protection regulations for startups. The legal framework for IPR protection under the Agreement on Trade-Related Aspects of Intellectual Property Rights needs to be revised. Viet Nam needs to implement the provisions of these agreements, and the government needs to make more efforts to raise awareness of the importance of IPR for startups.

Develop the legal framework for the establishment and operation of incubators and accelerators. The policies should also include capital financing mechanisms, financial incentives, and the creation of a capital mobilization mechanism for the establishment and operation of startup incubation centers.

Encourage joint activities between universities, research institutes and centers, and the private sector. Universities have the greatest potential for creating startups, as they themselves drive research projects for students and offer STEM courses. Encouraging joint activities between universities, institutes, research centers, and businesses to commercialize new scientific research will create incentives for more vibrant collaboration between research and business. The government can further support the exchange of ideas between domestic universities, research centers, and incubators with international partners.

Evaluate the activities of incubators, accelerators, and support centers at universities. An assessment must be made of what has been achieved and what is beyond the capacity of startups to propose solutions on how to improve their operations in the future. In this context, the connecting role of MOST in organizing seminars and forums is important.

3.2 Recommendations for an Enabling Ecosystem for Agritech and Healthtech Startups

Develop training programs and raise awareness of agritech startup development in the context of Viet Nam's ambitions for Industry 4.0. This could be achieved through exchanges between ICT departments and business, entrepreneurship, or management departments, as well as multidisciplinary courses to encourage students to combine the different fields to build successful innovation models.

Grant preferential loans and establish investment funds specifically targeted to agritech. The government could grant preferential investment policies or tax breaks to promising agricultural startups and issue a call for specific investments in the agritech startup ecosystem.

Support collaboration of agritech startups with domestic and foreign agribusinesses, organizations, and corporations to facilitate information sharing and identify needs and solutions. Since established large corporates in traditional industries recognize the need to adapt to new technology to remain competitive, they are potential customers, partners, and even VC investors for tech startups. Programs could be explored to organize the transfer of students and graduates to learn and train in developed countries where agritech is being developed.

In addition, it may be necessary to have policies to bring those who have completed their practical training, internships, and work abroad back to the country with new experiences and knowledge to work and build startups. Locally, the development of a national network for agritech startups to exchange ideas and organize meetings would be welcomed by agritech startups. This could also be linked to platforms for startups in other sectors.

Streamlining administrative procedures as a prerequisite for medical treatment is a general need for the health-care system. One step in this direction is Viet Nam's introduction of the national One Card program for health-care identification. In this regard, the program may need the support of the private sector to expand its services. Healthtech startups could support the government's attempt to digitize health-care services.

Encourage healthtech startups to collaborate with medical professionals, clinics, and hospitals. In the long term, medical providers will need the support of healthtech startups to streamline their operations, such as customer database, patient registration, appointment scheduling, and mobile payments.

Appendixes

Appendix 1: List of Laws Consulted

Government of Viet Nam. 2016. Decision No. 844/QD-TTg dated 18 May 2016 Approving the Project "Supporting the National Startup's Ecosystem until 2025". Ha Noi.

Government of Viet Nam. 2016. Decree No. 60/2015/ND-CP that Sets Foreign Ownership Limits in Vietnamese Enterprises.

Government of Viet Nam. 2016. Government Resolution No. 35/NQ-CP of 16 May 2016 on Supporting and Developing Enterprises till 2020. Ha Noi.

Government of Viet Nam. 2017. Law on Supporting Small and Medium Enterprises 2017. No. 04/2017/QH14. Ha Noi.

Government of Viet Nam. 2018. Decree No. 34/2018/ND-CP on the Establishment, Organization, and Operation of the Fund. Ha Noi.

Appendix 2: List of Startups Interviewed

NextFarm is an agritech startup that uses technology and automation to improve the quality of agricultural products. It is the first company in Viet Nam to combine a plant nutrition system with a meteorological monitoring solution. The company has developed sensors above and below ground and linked them to an app it also developed. The app provides farmers with information about soil health, such as salinity and fertility, as well as information, such as humidity and light intensity. Viettel Group selected NextFarm as a partner to implement the company's Smart Agriculture Solution program, which aims to revolutionize farming in Viet Nam and neighboring countries. NextFarm is a graduate of the Vietnam Silicon Valley Accelerator Program.

APPA Group is an agritech startup that combines the Internet of Things and big data. Its solutions include software and equipment that enable farmers to monitor the crop environment at any time of day and manage and operate production equipment from anywhere via smartphones and computers. APPA was the only agritech enterprise among the finalists in the 2018 Techfest Innovation Contest, and also received the runner-up award at the 2018 National Startup Contest. In 2019, APPA passed the standards of the Vietnam State Appraisal Council to be categorized as a science and technology enterprise. The council evaluates the quality of projects by large private companies and recommends them to the Prime Minister for approval. For smaller companies, the council can evaluate their projects and recommend that they be placed in a certain category that can bring tax or investment benefits.

SGS Smart Agriculture Solutions is engaged in the research and development of precision agriculture and soil fertility management. It offers modern techniques and comprehensive agricultural solutions for production. The company's smart soil technology solutions, particularly for soil classification and mapping, use drones, GPS positioning, high-resolution satellite imagery and mapping software to help farmers plan their fertilization and irrigation programs. As part of its soil sampling and analysis solutions, the company converts soil analysis data into practical fertilization programs and provides the farmer with an agronomist to help interpret the results.

Agri Hitech produces bio-organic fertilizer concentrated with high-quality microorganisms and processed using advanced technology to quickly and completely treat farm animal waste and cost-effectively eliminate odors and wastewater. Their range of bio-organic fertilizer production solutions uses superspeed fermentation technology that reduces composting time to 12 hours and eliminates the unpleasant odor of decomposing microorganisms.

MrFarm Agriculture 4.0 is an agritech startup that has been providing solutions to the larger and more established MrVina since 2017. The agritech startup's two products are Lean Smart Agriculture and MrFarm i4 Specialist, which solve problems of small and medium-sized farms with automated farming. MrFarm's Lean Smart Agriculture solutions use an automated irrigation and fertilization dosing system for farms between 2 and 3 hectares in size. The startup develops the technology and programs that connect climate and soil sensors to a control center, which in turn is connected to an irrigation and fertilization machine. Their product enables farmers to practice precision agriculture and save production costs.

One Vietnam Investment and Development Technology Joint Stock Company is a health technology startup developing electronic health records (EHRs) that allow patients to quickly connect with doctors, medical professionals, and specialized health-care facilities via their cell phones or iPad. Their product, EHRs Checkin, facilitates the admission (readmission) of patients to hospitals for reexamination or for consultation. Hundreds of companies, health-care organizations, and individuals use the application.

VietSoftware Technology Solution is a software outsourcing service provider that initially offered systems integration solutions for banking and finance. The company has dedicated a pool of its developers to build a smart health application that connects patients with doctors using a similar EHR system to One Vietnam Investment and Development Technology JSC.

Innovative Solution for Health Care (ISOFH) Technology Joint Stock Company provides digital information management solutions for hospitals. Their product called Overall Hospital Management Software System enables the integration of information from different hospital departments into one platform.

OneLink Viet Nam Company Limited is a startup company working with the Department of E-Commerce and IT, the Department of Medical Examination and Health Treatment, and the Ministry of Industry and Trade to implement the National One Card Program. The company is developing the app, kiosks, and digital solutions for the national "health smart card," which will enable online registration for medical exams and registration, exchange of medical information and patient medical history, and electronic payment to hospitals and health-care service providers. The smart health card is being used in more than 20 hospitals in 25 provinces and cities.

Finizz is a healthtech startup based in Ho Chi Minh City that provides an electronic platform for medical services. The app helps users find and book the most suitable doctors and private clinics for their medical needs (based on qualifications and distance). The platform allows patients to give reviews on the service quality of doctors and private clinics to help other patients choose health-care providers.

VieVie Healthcare is a startup that has developed a paid app that matches patients with doctors in real time. It has developed a platform that allows patients and doctors to have a conversation, and connect and share information. This app has increased the reach of medical services in the provinces.

References

Amway. 2018. *Amway Global Entrepreneurship Report: What Drives the Entrepreneurial Spirit?*.

Chung, S. Y., D. C. Lieu, N. T. Vinh, and D. Y. Han. 2015. *Venture Capital Investment in Vietnam: Market and Regulatory Overview*. Baker and McKenzie (Vietnam) Ltd.

Das, K. 2019. *Investments in Vietnamese Startups Tripled in 2018*. Vietnam Briefing.

DO Ventures. 2020. Vietnam Tech Investment Report 2019 – H1/2020.

Global Entrepreneurship Monitor (GEM) and Vietnam Chamber of Commerce and Industry (VCCI). 2018. *Global Entrepreneurship Monitor Vietnam Report 2017/2018*. Thanh Nien Publishing House.

Government of Viet Nam. 2017. Law on Supporting Small and Medium Enterprises 2017. Article 3. No. 04/2017 / QH14. Ha Noi.

Namjatturas, J. 2018. *Vietnam's Startup Ecosystem: All You Need to Know*. Techsauce Summit.

National Innovation Center (NIC). 2021. Vietnam Innovation and Tech Investment Report 2021. https://doventures.vc/en/insights/reports/vietnam-innovation-and-tech-investment-report-fy2021.

Nguyen, M-N. 2021. *Number of Investment Deals in Tech Startups in Vietnam from 2013 to the First Half of 2020*. Statista.com (in Vietnamese).

Nguyen, N. 2022. Capture Vietnam's Investment Landscape in 2021. VICGO. https://vicgo.co/2022/03/vietnams-investment-landscape-2021/.

Nguyen, P. Q. and N. V. Le. 2020. *Vietnam's Startup Ecosystem (3): Non-Government Support for Startups—Who Is Taking Part in the Game?*. Vietnam UpStars Team.

Organisation for Economic Co-operation and Development (OECD). 2015. *Cross-Country Evidence on Start-Up Dynamics*. In F. Calvino, C. Criscuolo, and C. Menon, eds. *OECD Science, Technology and Industry Working Papers*. No. 2015/06. Paris: OECD Publishing.

Phan, T. 2021. Startup Ecosystem in Vietnam. Bachelors Thesis (in Vietnamese). University of Applied Sciences, Vaasan Ammattikorkoeakoulu.

Quynh, N. 2018. Project 844: A Catalyst for Vietnamese Startup Surge. *Vietnam Economic News*. 26 October.

Thong, V. 2021. Vietnamese Startups Draw $1.3 billion Worth of Investments in 2021. VN Express International.

Topica Founder Institute. 2017. 2016 Startup Deals Vietnam. https://www.slideshare.net/topicafounderinstitute/vietnam-startup-deals-insight-2017-87618940.

United States Department of State. 2016. American Innovation Roadshow: Senior Advisor Thorne Travel to Indonesia, Vietnam, and the Philippines. Media Note. Washington, DC.

Van Roosmalen, M. and D. Totten. 2019. *Country Report: Vietnam – Entrepreneurial Ecosystem Assessment*. Dutch Good Growth Fund. Ministry of Foreign Affairs and Emerging Markets Consulting.

Vietnam Advisors. n.d. Eight Funds Investing in Vietnamese Startups.

Vietnam Chamber of Commerce and Industry (VCCI). 2017. Mechanism Supporting Creative Enterprises (in Vietnamese).

Vietnam Insider. 2019. Each Vietnamese Citizen Will Be Issued with a Unique Health ID Card.

Vietnam Insider. 2020. Vingroup Restructures a Startup Venture into Industrial Real Estate Development.

Viet Nam News. 2018. Vietnam Wants to Nurture Entrepreneurship among Kids.

Viet Nam News. 2019. PM Welcomes SoftBank's Investment in Viet Nam.

VietnamPlus. 2017. Project Support Women's Startup.

World Economic Forum (WEF). 2014. *Entrepreneurial Ecosystems Around the Globe and Early-Stage Company Growth Dynamics – The Entrepreneur's Perspective*. Geneva.

CPSIA information can be obtained
at www.ICGtesting.com
Printed in the USA
LVHW071910111122
732929LV00009B/328

9 789292 696963

Viet Nam's Ecosystem for Technology Startups

Technology-based startup enterprises are an increasingly important part of the business landscape in Asia and the Pacific. By applying innovative technologies to create new products and services, they can make a significant contribution to economic development while generating social and environmental benefits. However, to survive and then thrive, tech startups require an enabling ecosystem that includes supportive government policy, adequate access to capital, skilled personnel, and quality digital infrastructure. This report examines Viet Nam's innovative enterprises in two sectors: agriculture and health. It identifies challenges in their quest to scale up and offers practical recommendations to overcome these challenges and create an enabling ecosystem in which startups can grow.

About the Asian Development Bank

ADB is committed to achieving a prosperous, inclusive, resilient, and sustainable Asia and the Pacific, while sustaining its efforts to eradicate extreme poverty. Established in 1966, it is owned by 68 members —49 from the region. Its main instruments for helping its developing member countries are policy dialogue, loans, equity investments, guarantees, grants, and technical assistance.

ISBN 978-92-9269-630

9 789292 696306

ADB

ASIAN DEVELOPMENT BANK
6 ADB Avenue, Mandaluyong City
1550 Metro Manila, Philippines
www.adb.org